I0469353

Tire Recycling Facility Fire
Nebraska City, Nebraska

Investigated by: John Lee Cook, Jr.

This is Report 145 of the Major Fires Investigation Project conducted by Varley-Campbell and Associates, Inc./TriData Corporation under contract EME-97-CO-0506 to the United States Fire Administration, Federal Emergency Management Agency.

Department of Homeland Security
United States Fire Administration
National Fire Data Center

U.S. Fire Administration Fire Investigations Program

The U.S. Fire Administration develops reports on selected major fires throughout the country. The fires usually involve multiple deaths or a large loss of property. But the primary criterion for deciding to do a report is whether it will result in significant "lessons learned." In some cases these lessons bring to light new knowledge about fire--the effect of building construction or contents, human behavior in fire, etc. In other cases, the lessons are not new but are serious enough to highlight once again, with yet another fire tragedy report. In some cases, special reports are developed to discuss events, drills, or new technologies which are of interest to the fire service.

The reports are sent to fire magazines and are distributed at National and Regional fire meetings. The International Association of Fire Chiefs assists the USFA in disseminating the findings throughout the fire service. On a continuing basis the reports are available on request from the USFA; announcements of their availability are published widely in fire journals and newsletters.

This body of work provides detailed information on the nature of the fire problem for policymakers who must decide on allocations of resources between fire and other pressing problems, and within the fire service to improve codes and code enforcement, training, public fire education, building technology, and other related areas.

The Fire Administration, which has no regulatory authority, sends an experienced fire investigator into a community after a major incident only after having conferred with the local fire authorities to insure that the assistance and presence of the USFA would be supportive and would in no way interfere with any review of the incident they are themselves conducting. The intent is not to arrive during the event or even immediately after, but rather after the dust settles, so that a complete and objective review of all the important aspects of the incident can be made. Local authorities review the USFA's report while it is in draft. The USFA investigator or team is available to local authorities should they wish to request technical assistance for their own investigation.

This report and its recommendations was developed by USFA staff and by Varley- Campbell and Associations, Incorporated (Miami and Chicago), its staff and consultants, who are under contract to assist the Fire Administration in carrying out the Fire Reports Program.

The Federal Emergency Management Agency, United States Fire Administration gratefully acknowledges the cooperation of the Mayor and members of the Nebraska City Fire Department, the Nebraska City Rescue Squad, the Nebraska State Fire Marshal's Office, Williams Fire and Hazard Control, Inc., and Region VII of the United States Environmental Protection Agency Emergency.) Every one who assisted in the preparation of this report was generous with his or her time, expertise, and counsel.

For additional copies of this report write to the United States Fire Administration, 16825 South Seton Avenue, Emmitsburg, Maryland 21727. The report and the photographs, in color, are available on the Administration's Web site at http://www.usfa.dhs.gov/

U.S. Fire Administration

Mission Statement

As an entity of the Department of Homeland Security, the mission of the USFA is to reduce life and economic losses due to fire and related emergencies, through leadership, advocacy, coordination, and support. We serve the Nation independently, in coordination with other Federal agencies, and in partnership with fire protection and emergency service communities. With a commitment to excellence, we provide public education, training, technology, and data initiatives.

TABLE OF CONTENTS

Tire Recycling Facility Fire
Nebraska City, Nebraska
January 23 to February 3, 2002

Investigated By: John Lee Cook, Jr.

Local Contacts:

The Honorable Jo Dee Adelung, Mayor
City of Nebraska City
1409 Central Avenue
Nebraska City, NE 68410
(402) 873-6080

Alan Viox, Fire Chief
Nebraska City Fire Department
1409 Central Avenue
Nebraska City, NE 68410
(402) 873-6293

Larry L. Wiles, Historian
Nebraska City Fire Department
Local Contacts: Susan McGown, Chief
Nebraska City Rescue Squad
1409 Central Avenue
Nebraska City, NE 68410
(402) 873-3422

Daniel Kelly, President and CEO
St. Mary's Hospital
1314 North 3rd Avenue
Nebraska City, NE 68410
(402) 873-8901

Ken Winters, Nebraska State Fire Marshal
246 South 14th Street
Lincoln, NE 68508
(402) 471-2027
Janice Kroone, On-Scene Coordinator
U.S. Environmental Protection Agency Region 7
901 North Fifth Street
Kansas City, KS 66101
(913) 551-7005

Chancey Naylor
Williams Fire and Hazard Control
PO Box 1359
Mauriceville, Texas 77626
(800) 231-4613

OVERVIEW

The members of the Nebraska City Fire Department were dispatched to a reported chimney fire at a housing complex for the elderly at 02:54 hours on the morning of Wednesday January 23, 2002. The first engine company to arrive on the scene discovered that the fire was actually located a few blocks away at the EnTire Recycling Center. The facility chips up old tires and produces a raw product that is used to manufacture synthetic athletic turf and playground surfaces.

Firefighters discovered a working fire at the facility that had spread to one of the production buildings as well as to the raw product in several of the silos at the site. The complex had formerly been used as a grain facility and product was stored in varying amounts in five metal silos originally constructed to store grain. Most of the fire was extinguished, but the fires in the silos were allowed to continue to burn because of poor visibility and the potential for collapse. Mutual aid was immediately summoned to assist in the extinguishment effort.

Cooling water was directed on the silos and liquid nitrogen was used in an attempt to smother the fire. An explosion occurred during the afternoon of the first day and injured thirteen firefighters. Four of the firefighters sustained injuries serious enough to require hospitalization. Fears of a subsequent explosion and a continuing possibility of collapse resulted in the firefighters adopting a defensive mode of operations.

The fire was not fully extinguished until February 3, 2002 and required the efforts of emergency responders from over thirty fire and EMS agencies. Final extinguishment was accomplished by Williams Fire and Hazard Control, a private contractor renowned for the extinguishment of oil well and flammable liquid tanks fires. The firm was hired by the Environmental Protection Agency.

The fire posed a significant environmental threat due its proximity to the Missouri River and the toxic byproducts of the burning tires. State and Federal environmental officials monitored the extinguishment effort throughout the incident. During the incident, portions of the community had to be evacuated on two occasions due to a potential for additional explosions and the toxicity of the smoke.

KEY ISSUES

Issue	Comments
Codes	While the city had adopted the Uniform Building and Fire Codes, neither local nor State codes prohibited or regulated a facility of this type from locating and operating within the city. Strict zoning and building codes are necessary to prohibit such operations. The facility was not originally designed to accommodate the product and operations being conducted at the site.
Duration of Incident	The incident lasted for eleven days, seven hours, and fifty-six minutes. Very few agencies have the staffing and resources to efficiently and effectively manage an incident of this duration. Members of volunteer and combination departments have employment and family issues that are impacted by an incident of this duration. Reliance upon mutual aid companies and a rotating fire watch allowed the department to successfully manage the incident.
Structural Integrity	Structural collapse posses a significant risk in an incident of this type. The silos were constructed of unprotected steel and were not designed for direct flame impingement or for containing a deep-seated fire. The Otoe County Sheriffs Office used crime scene/vehicle collision investigation equipment to monitor the movement in the silos to warn of potential collapse.
Environmental Damage	The incident was located immediately adjacent to the Missouri River and the run-off from the extinguishment effort posed a significant risk to the environment. Throughout the incident, State and Federal environmental agencies monitored air and water quality as well as assisted with diking and containment efforts. Approximately 390,200 gallons of contaminated water and 3.3 tons of tire crumb material were removed from the site.
Community Support	There was a tremendous show of support for the department throughout the incident. Citizens provided food and coffee to the emergency responders.
Communications	The fire department operated on a single radio frequency that failed during the incident and there was an interoperability issue with interoperability with mutual aid companies from Iowa.
Shelter and Evacuation Plans	It became necessary to evacuate large portions of the community on two occasions because of the potential for another explosion and the toxicity of the products of combustion. The impacted area included a elderly housing unit with seventy-nine residents as well as several schools. Emergency Operations Plans must address the process of evacuating large areas, transportation of special populations, as well as the establishment of shelters.

THE COMMUNITY

Nebraska City is the seat of Otoe County and is located in the southeastern portion of Nebraska approximately forty-five minutes south of Omaha and forty-five minutes east of Lincoln. The community of 7,200 residents is located on the west bank of the Missouri River adjacent to Iowa. Founded as a river port, the community was surveyed in 1854 and incorporated in 1856.

Fire protection is provided by the Nebraska City Volunteer Fire Department, which is commanded by a cadre of three chief officers and three captains. Established in 1856, the department was the first fire department organized in Nebraska and is composed of forty- five volunteer members and three career drivers that work a 24/48-shift schedule. The department operates two engines, an aerial tower, a heavy rescue, and a mobile command post, which are all housed in a single station located downtown. The station opened in 1972.

Also housed at the station is the Nebraska City Rural Fire District, which is a separately chartered and funded agency. The rural department is, however, staffed by many of the same personnel and its apparatus fleet includes two tankers, two brush units, and an engine.

Emergency Medical Services are provided by the Nebraska City Rescue Squad, which is an all volunteer agency with thirty-eight members. The Rescue Squad provides transport services at both the basic and advanced life support level. The squad is quartered at the fire station.

THE FACILITY

The fire and explosion occurred at a facility owned and operated by EnTire Recycling, Incorporated. Founded in 1995, the corporation had a workforce of fifteen when the fire occurred and operated under a permit issued by the Nebraska Department of Environmental Quality as a scrap tire collector, processor, and hauler. EnTire recycled old tires into reusable rubber for products such as athletic tracks, artificial turf for sports stadiums, and playground covers.

The plant (Elevation 939 feet) was located on the eastern edge of the community at 215 North First Street between the mainline of the Union Pacific Rail Road and the Missouri River (Elevation 917 feet), some 200 yards to the east. Until the 1960s, the facility had served as a grain operation. The facility consisted of two 100 ft. x 300 ft. production buildings, an office, four silos that were 64 feet high and twenty-eight feet in diameter, a shorter corrugated silo (see diagrams and photos), and two 8,000-gallon liquid nitrogen tanks. The nitrogen tanks had been filled just prior to the incident according to a company spokesperson. The facility was on land owned by the Brock Grain Company of Brock, Nebraska.

Bias and steel belted tires were stored in the south building where they were shredded into chips (crumb rubber) of approximately two to four inches in diameter (see photo). The chips were then moved by conveyor into the silos for storage. In the north building, the chips were frozen by nitrogen and were ground into fine particles in a process that involved separating the metal particles from the rubber.

Each silo held approximately 100,000 shredded tires with an estimated weight of 1,000 tons. At the time of the fire, the northwest silo was one-half full, the northeast silo was three-quarters full, the southwest silo was completely full, and the southeast silo was full to the sixteen-foot level. The short, corrugated silo was empty.

The plant had been inspected by Nebraska State Fire Marshal's Office on more than one occasion and had a history of poor housekeeping as well as a track record of poor equipment maintenance. There had previously been two fires at the site, as well as two fires at another EnTire facility, also located in Nebraska City.

THE INCIDENT

Upon receipt of a 9-1-1 call, the Nebraska City Volunteer Fire Department and the Rescue Squad were dispatched at 02:54 hours on the morning of Wednesday, January 23, 2002 to a report of a chimney fire at the Riverside Terrace Building, a six story residential housing unit for the elderly. The County Sheriffs Office answers all 9-1-1 calls and dispatches for the fire department and rescue squad. A base station radio is maintained at the fire station and in the event of a working fire, a firefighter operates the station radio and handles all of the radio traffic generated by the incident.

A standard first alarm assignment consisting of two engines, a truck, a heavy rescue, and an ambulance responded to the incident staffed by twenty-seven firefighters and two emergency medical technicians. The first engine company on the scene determined that the fire was actually at the EnTire Recycling Plant, located several blocks east of Riverside Terrace.

Other responding companies were diverted and upon arrival discovered fire burning in the four tall silos, on the ground between the silos, and in the south production building. Firefighters also observed that the southwest silo was listing to the east. They became immediately concerned about the potential for collapse because the silos were constructed of unprotected steel and it was unclear how long the fire had actually been burning prior to their arrival.

A second alarm was dispatched, which summoned another Nebraska City engine to the scene as well as two engines each from Dunbar, Plattsmouth, and Syracuse. Plattsmouth also responded with a truck company.

Approximately fifty-five firefighters began to extinguish the exterior fire between the silos and in the south production building. Their efforts took one and one-half hours, but they elected to wait until daylight before attempting to extinguish the fire in the silos because of limited visibility and the potential for collapse. Given the magnitude of the incident, officials made a decision at 03:15 hours to shut down the rail traffic on the Union Pacific mainline, which was located immediately adjacent to the plant, to the west.

At the time of the alarm, the temperature was below freezing. Temperatures on the January 23rd ranged from a low of 26 degrees Fahrenheit to a high of 50 degrees Fahrenheit, wind chill was 11 degrees Fahrenheit, and winds were from the northwest at 11 to 19 mph with gusts to 26 mph. There were scattered clouds, but no precipitation.

By approximately 10:45 hours, mutual aid companies were released as well as all the majority of the Nebraska City companies. An engine company, a rescue squad, an ambulance, and twenty-five firefighters remained at the scene, primarily performing overhaul operations. No aggressive fire suppression effort was attempted, as command personnel tried to determine how best to extinguish the deep-seated fire in the silos.

After consultation with the State Fire Marshal's Office, nitrogen was applied in an attempt to extinguish the fire by depriving it of oxygen, a technique that had been used successfully at grain elevator fires within the State. The nitrogen was obtained from two tanks located at the facility. At 12:30

hours, firefighters noted that the southwest silo had begun to lean into the southeast silo and that the head house located on top of the four silos had begun to shift to the south.

At 13:13 hours, while the nitrogen was still being applied an explosion occurred and injured thirteen firefighters. Fortunately, the injuries proved not to be life threatening. The cool weather had resulted in the firefighters remaining full bunkered out and personnel engaged in firefighting activities were wearing SCBAs. Reports from the scene indicate that the fire blew out twice from the silo like a torch just prior to the explosion. An evacuation order had been given and the explosion occurred as firefighters were beginning to leave the area.

The explosion expelled chunks of rubber that were approximately two to four inches in diameter. Debris was hurled 300 to 400 feet away, damaging and destroying apparatus and equipment as well as injuring the thirteen firefighters. A list of the destroyed equipment is included in Appendix C.

After the explosion, Command immediately requested additional EMS units. When dispatched, the EMS units were notified that there were causalities. Responding crews erroneously believed that there were fatalities involved, which was unsettling since Nebraska City is a small, tight-knit community. They also were unaware of exactly what had exploded, which magnified their apprehensions since they did not know what they would encounter at the incident site.

The Rescue Squad's Captain assumed command of EMS operations and established a command post on the south side of the fire. An EMS staging area was established on the north side and soon seventeen ambulances were in queue in response to a radio broadcast for all available ambulances to respond. Three ambulances were placed in the primary staging area and two units were maintained in a secondary staging point, while others transported the injured firefighters to the hospital. A medical transport helicopter was also dispatched to the scene from Lincoln in case injures proved to be of a serious nature.

EMS crews immediately triaged all the firefighters at scene. The thirteen with injuries were transported to St. Mary's Hospital, including eleven by the local rescue squad. The three most severely injured were transported by the ambulance that was on standby at the scene when the explosion occurred. One had a broken leg and two were completely coated in heavy black residue. Three of the victims were admitted to the local hospital and one patient was transferred to Bryan LGH West Medical Center in Lincoln.

Injuries included one compound fracture of the left leg; one fracture of the right ankle; one left ankle injury; numerous contusions, scraps, shrapnel wounds, smoke inhalation, respiratory injuries, and cuts. There were no burn injures. Fire department officials estimated that the cost of medical treatment was approximately $13,406.

In a third wave of triage, EMS personnel found three firefighters that needed to take their medications, which had been ignored as a result of the fire. They were relieved of duty until they had been properly medicated.

St. Mary's, the local hospital, instituted its emergency plan, which included the recall of off-duty physicians and prepared to treat the victims. Omaha Methodist Hospital contacted St. Mary's and told them they were on standby in the event that they were needed.

As previously noted, the Nebraska City Fire Department and the Rescue Squad are both dispatched by the Otoe County Sheriffs Department. During an incident, a firefighter reports to the fire station and assumes the communications duties for the incident using a base station located in the watch office.

A single, low-band VHF frequency is used for this purpose. The department also monitors fire traffic in Iowa, which is just across the Missouri River from the City.

However, Iowa departments operate on high band and cannot directly communicate with units in Nebraska.

During the height of the activity surrounding the explosion, a microphone got stuck in the open position and disrupted all radio communications. For all practical purposes, this left companies at the scene without any communications other than by cellular telephone.

Following the explosion, the Nebraska City apparatus previously released were recalled to the scene. Two staging areas were established, one at the fire station and another in the residential area near the incident. Including mutual aid resources, a total of nine engines, two trucks, two tankers, three rescues, and twelve ambulances were used throughout the suppression effort and subsequent operations.

The initial tactical objective was to keep the silos cool with hose lines and portable deluge sets. At one point firefighters had just walked out from the center of the silos when fire and debris fell from the top and landed on the spot that they had been just moments before. The owner removed some of the materials with a skid loader, as had been done at prior calls at the facility.

Following the explosion, a command post was established in the municipal power plant located 600 feet north of the fire. The facility offered a safe; climate controlled environment as well as an unobstructed view of the incident site. An outer perimeter was established to limit access to the site and public works and parks personnel set up barricades.

At 11:30 hours on January 25th, fire spread from the conveyor to the north building when the wind direction changed, requiring firefighters to evacuate the building. A rumbling sound was heard, prompting the fear of a potential backdraft explosion. The wind also caused smoke to drift into a populated area of the city, which included several schools and the Riverside Terrace facility. As a precaution, Command ordered an evacuation of a 30-block area west of the fire. Residents were allowed to return to their homes at 15:30 hours when the fear of the explosion had subsided and the wind once again changed directions.

Mutual aid was called a second time at 11:43 hours on January 27th when fire erupted again in the north production building. Companies from Dunbar, Syracuse, Hamburg, and Sidney responded. A second evacuation order was issued, but was rescinded when the fire was contained by mid-afternoon.

A Public Information Officer operated from the fire station and gave periodic updates throughout the duration of the incident. All of the major television networks contacted the department as well as the local television stations from Lincoln and Omaha. Radio stations and newspapers in a four State radius also contacted the department as well as Firehouse Magazine.

Firefighters estimated that eleven (11) million gallons of water were used to extinguish the fire. Water supply was adequate throughout the entire incident, in part because personnel from the utilities department responded early in the incident and ran pumps as needed and kept water storage facilities full. The city's water supply system has a capacity of seven million gallons a day and is totally supplied from wells.

High-tension electric lines were located immediately east of the plant. Firefighters were concerned that the fire might damage the wires or that a collapse would bring the wires down further complicating matters. The power was shut off and an attempt was made to relocate the lines. The effort was abandoned on January 26th due to a further shift of the silos to the east. The risk to the linemen working in bucket trucks was deemed to be too great.

Local law enforcement personnel established a perimeter and were used to handle traffic and to keep unnecessary people out of the area. They also assisted with the two evacuation efforts on the 25th and 27th of January. The local radio station assisted with evacuation efforts and provided citizens with periodic updates throughout the incident.

The prolonged extinguishment effort necessitated the establishment of a rotating work schedule. At night, most operations were suspended, a fire watch was maintained to insure that the fire did not gain headway and spread to nearby exposures. Mutual aid department assisted in this effort. See Appendix B for a copy of the rotation schedule.

The city experiences another working structure fire on January 30th while the incident was still ongoing. At 07:26 hours, a fire was dispatched at the Wurtele Building, southwest of town. The fire was declared under control at 08:26 hours. Thirty-seven firefighters from nineteen different communities responded to the incident with three engines, one truck, twenty-one tankers, one rescue squad, and an ambulance. The water supply is very poor in that part of the community, thus the large number of tankers that responded.

Figure One: Sequence of Events on the First Day--January 23, 2002

Time	Event
02:54	Alarm dispatched
03:15	Rail traffic is halted on the Union Pacific mainline located on the west side of the plant
04:24	The exterior fire and the fire in the south building are extinguished and firefighters decide to wait until daylight to extinguish the fire in the silo
10:45	Mutual-aid companies released, believed to be in the overhaul phase
12:30	Southwest silo begins to lean into the southeast silo and the head house begins to lean into scaffolding
13:13	Explosion in SE silo occurs injuring thirteen firefighters; requested mutual aid for pumpers and ambulances
13:28	Microphone stuck in an open position disabling the fireground radio frequency
13:39	The first three injured firefighters transported to the hospital, followed by four more at 13:42

Figure Two: Daily Chronology of Events

Date:	Event
23 January	02:54 received a 9-1-1 call reporting the fire. Dispatched as a chimney fire at Riverside Terrace, a six-story apartment building for senior citizens. Mutual aid was requested and firefighters extinguished the exterior fire and the fire in the south building.
	An explosion at 13:13 hours sent thirteen firefighters to the hospital.
24 January	The Red Cross opened shelters and fifty-eight residents from Riverside Terrace were taken to the First Methodist Church.
25 January	The State Fire Marshal contacted the Department of Environmental Quality who in turn contacted the EPA and the Coast Guard Hazmat task force. Together with the insurance company debated whether to let it burn or extinguish the fire.
	Fire spread from the conveyor to the north building at 11:30 hours when the wind direction changed, requiring evacuation of the building. A rumbling sound is heard, prompting fear of a potential backdraft.
	A 30-block area west of the fire was evacuated until 15:30 hours.
26 January	The fire continues to smolder in the two southern silos.
	EnTire's insurance company hired Williams Fire and Hazard Control for consultation. They arrived at 13:10 hours and held a strategy meeting at 14:00 hours. The insurance company decided it would cost too much to extinguish the fire and turned the operation over to the EPA. EPA's ERRS hired Williams to extinguish the fire in the silos.
	Weather: Low 32 degrees Fahrenheit, High 68 degrees Fahrenheit, wind chill 24 degrees Fahrenheit; cloudy, wind NW 7-10 mph.
27 January	Silos shift and the head house lists even more.
	Train traffic was halted due to observations that the silos listed more to the east when a train passed by even at reduced speed.
	Mutual aid called a second time at 11:43 hours from Dunbar, Syracuse, Hamburg, and Sidney when fire erupts in the north building. A second evacuation order is issued and the fire is contained by mid-afternoon.
28 January	Two of the silos collapsed. The head house fell at 08:00 and the southeastern and southwestern silos collapse at 10:22 hours.
	Fire erupts and efforts are directed at preventing the fire from spreading into the south building.
	Weather: Low 17 degrees Fahrenheit, High 44 degrees Fahrenheit, Wind chill 0 to 29 degrees Fahrenheit; overcast with winds from the NW at 12 to 18 mph with gusts to 27 mph.
29 January	Began the process of taking the silos apart and extinguishing the burning materials.
30 January	Continued process of taking the silos apart and extinguishing the burning materials.
	07:26 hours, a fire occurred at the Wurtele Building, southwest of town. Fire declared under control at 08:26 hours, but required the efforts of three engines, one truck, twenty-one tankers, one rescue squad, and an ambulance and the efforts of thirty-seven firefighters from nineteen communities to extinguish.
31 January	Completed the demolition of the silos.
	Weather: Overcast with snow, winds from the west 14 to 18 mph. Temperature range: 21- 26 degrees Fahrenheit, wind chill 4-11 degrees Fahrenheit.

Date:	Event
1 February	Fire is turned over to Williams Fire and Hazard Control and the local fire department is no longer an active participant. Extinguishment efforts are carried out during the day and at night, only a fire watch is present.
	All material from the southeast and southwest silos had been removed, extinguished, and transported to the landfill.
	Weather: Cloudy with eight inches of snow. Temperatures range from 1 to 25 degrees Fahrenheit, with wind chills from -24 to 9 degrees Fahrenheit. Wind gusts to 25 mph.
2 February	Rubber is removed, wet down, and trucked to a landfill in Butler County.
	A giant "can opener" is used to rip apart remaining silos and the fire in the northwest silo was extinguished.
3 February	At 10:50 hours, the fire was declared out some eleven days, seven hours, and fifty-six minutes after initial alarm.

EXTINGUISHMENT

On January 25th, EnTire's insurance company agreed to bring in two outside consultants from a commercial firefighting company to assess the situation. They arrived on the afternoon of the 26th. The company did not, however, work for the EPA's Emergency Response and Removal contractor so procedural and contractual delays prevented the company from taking any immediate action. Those issues were soon resolved and the consultants from this company ordered their equipment and personnel to respond to the incident.

Their personnel arrived on January 28th, and on January 29th firefighters began the process of taking the remaining silos and debris apart. They removed burning materials, spread them out, and extinguished the burning rubber. This continued until January 31st. A track hoe was used to remove the east side of the south production building in order to gain access to the interior contents. The raw, unburned tires were removed from the south building and relocated elsewhere. Contractors built berms to contain runoff, cut-up the silos, hauled the extinguished debris to cool down areas and wet down the burning materials. Wreckers were used to pull steel debris from pile.

On February 1, 2002 the task of extinguishing the fire was turned over to the vendor and the Nebraska City Fire Department stood down. The firm specializes in shipboard and industrial fire control. At that point the fire department stood down.

The vendor brought the following equipment and personnel to the scene:

- One 2,500 gpm and one 3,000 gpm trailer mounted fire pumps
- 3,000 feet of five-inch hose
- 3,000 feet of three-inch hose
- 3,000 feet of 1-3/4-inch hose
- Five-inch monitor with a 2,000 gpm foam nozzle
- Six 2-1/2-inch monitors with foam nozzles
- Used 3,200 gallons of foam
- Eight firefighters

Personnel from the vendor essentially operated during daylight hours and shut down their operations at dark. A fire watch was maintained throughout the night to prevent a major rekindle or spread of burning materials. Freezing weather required the pumps to be enclosed by a tarp and the use of a kerosene space heater to maintain the flow of water and foam.

The fire was declared extinguished at 10:50 hours on February 3rd, some eleven days, seven hours, and fifty-six minutes after the dispatch of the first alarm. The fire took almost 272 hours to extinguish. According to Nebraska City utilities department, 11,390,000 gallons of water were used in the extinguishment effort. The contractor also used 1,600 gallons of foam. Approximately, 3,280 tons of waste tire crumb material was hauled to the Butler County Landfill in nearby David City, Nebraska. The EPA and the Coast Guard estimate that cleanup costs reached $1.4 million.

The fire department estimates that Nebraska City spent 2,542 man hours in their effort to contain the fire, while mutual aid agencies worked another 2,373 man hours. Equipment hours totaled 1,314.

ENVIRONMENTAL IMPACT

The close proximity of the incident site to the Missouri River created the potential for environmental damage due to run off from the water being applied to extinguish the fire. According to the EPA, burning tires produce pyrolithic oil that contains naphthalene, trichloroethane, tetrachloroethane, ethylene, toluene, poly aromatic hydrocarbons (PAHs), and heavy metals. Data obtained at other tire fires by the EPA indicate the presence in the air of the following contaminants: benzene, PAHs, phosgene, naphthalene, toluene, styrene, acrylonitrile, formaldehyde, carbondisulfide, sulfuric dioxide, carbon dioxide, and numerous heavy metals.

City and county public works personnel and equipment were used to transport sand to the incident site and where they constructed dikes and containment structures to prevent runoff from entering the river. Two collection points served to recover water, one each on the north and east sides of the fire. The EPA responded to the scene and their Emergency Response and Removal (ERRS) contractor collected runoff water, removed the oil, and disposed of the material. During the duration of the incident, a private contractor transported approximately 390,200 gallons of contaminated water by truck to the Omaha Waste Water Treatment plant for purification.

The number of contractors on site and the weariness of the local fire department, which had been severely taxed due to the duration of the incident, and loss of equipment and gear, resulted in the EPA requesting the assistance of the Coast Guard. The Atlantic Strike Team responded to provide assistance in documenting costs.

INVESTIGATION

The incident was investigated by the Nebraska State Fire Marshal's Office with the assistance of private investigators from EnTire's insurance company. Investigators secured the services of a metallurgist to examine whether the heat patterns on the steel point plates of the silo indicated an outside or inside heat source. The cause of the fire and explosion remains under investigation.

The State Fire Marshal has determined that the probable cause was a faulty bearing in the stiff leg auger which carried the crumb rubber from the ground level to the top of the silos. The bearing overheated igniting the rubber debris around the sugar base. A contributing factor was the poor housekeeping practices. No definite cause of the explosion has been identified, although the State Fire Marshal has theorized that when the liquid nitrogen was introduced into the space above the crumb rubber it froze the top several inches of rubber, sealing the top of the silo. The continued application of nitrogen in to the silo built up pressure. As firefighters and plant personnel worked to remove the burning rubber, a rupture in this seal led to an explosion.

LESSONS LEARNED

1. **Small towns can muster the resources to handle a major incident.**

   A considerable effort was expended by members of the local fire department as well as its neighbors in their efforts to manage and extinguish the incident. Agencies worked well together and relied upon the local emergency operations plan for guidance. The use of a fire watch was a very proactive step and helped minimize fatigue that is common at incidents of long duration. The overwhelming majority of the emergency responders were volunteer, which necessitated a great deal of time away from their families and jobs. Emergency planners should consider this possibility when preparing or revising their plans.

   Functional emergency operations plans provide for the appropriate division of labor prior to an incident and take the guesswork out of the details. People and agencies know what is expected of them and can readily accomplish their assigned tasks with little, if any direction.

2. **Most local zoning and building codes do not adequately address facilities of this type.**

   Traditionally, volunteer fire departments do not become involved in local zoning issues and often lack the resources to properly preplan target hazards. The facility in this incident was not designed for its use, nor were there adequate safe guards built into the process since the complex already existed and was modified to accommodate this particular process. Fire departments should become more proactive in the permitting and zoning process and should endeavor to ensure appropriate code enforcements in facilities of this type, whenever it is possible to do so.

3. **EMS standbys are tedious.**

   In the vast majority of all incidents, EMS standbys are routine, non-events. They are tedious at best, but most are boring. It is easy to become complacent and forget that something real can occur. The local rescue squad was, in this instance, mobilized from a routine situation to a multiple casualty event with thirteen injured and the potential for many more.

4. **Documentation is vital.**

   An excellent documentation process was followed by the local fire and EMS agencies. Such record keeping is essential for effective and efficient operations. Also, if the incident is ever declared to be a disaster and reimbursement efforts are necessary. Proper documentation also establishes historical records in the event of future litigation and provides a baseline in the event that there are any long-term health care problems involving emergency responders.

5. **Communications continues to be an issue.**

   Inadequate communications systems and equipment as well as the incompatibility of neighboring jurisdictions continues to plague emergency responders as they respond to major events. The radio systems in Iowa and Nebraska were incompatible and an equipment failure during the most critical portion of the incident essentially shut down the fire department's radio system. More planning and resources need to be given to updating communication systems and ensuring their interoperability with mutual aid assets.

APPENDICES

APPENDIX A

Agencies Involved In Response and Recovery

Photo	Description
Nebraska Mutual Aid Fire & EMS	Auburn, Bellevue, Brock, Burr, Cook, Douglas, Dunbar, Farraquat, Johnson, Julian, LaVista, Murray, Nehawka, Nemaha, Otoe, Palmyra, Peru, Plattsmouth, Southeast RFD, Southwest RFD, Syracuse, Tecumseh, Unadilla, Union
Iowa Mutual Aid Fire & EMS	Hamburg, Percival, Riverton, Sidney, Shenandoah, Tabor
Law Enforcement	Otoe County Sheriffs Office
State of Nebraska	Nebraska Department of Environmental Quality, Nebraska State Fire Marshal's Office
Federal Agencies	U.S. Environmental Protection Agency, U.S. Coast Guard
Other	Critical Incident Stress Management Team, Saint Mary's Hospital, First Methodist Church (shelter), Nebraska School for the Visually Handicapped (shelter), Otoe County Emergency Management Agency, Nebraska City Public Works and Utility Departments, Otoe County Public Works Department, Otoe County Chapter of the American Red Cross

APPENDIX B

Fire Watch Duty Schedule

Average of four firefighters on scene for each six-hour shift, so NC FF could rest

Date	00:00 to 06:00	18:00 to 24:00
1/23	Fire reported at 02:54	Nebraska City
1/24	Syracuse	Sidney
1/25	Nebraska City	Dunbar
1/26	Unadilla	Dunbar
1/27	Murray	Sidney
1/28	Nehawka	Otoe
1/29	Auburn	Palymyra
1/30	Tecumseh	Dunbar
1/31	Brock, Johnson, Nemaha	Shenadoah, Williams Fire
2/1	Nebraska City	

APPENDIX C

List of Destroyed Equipment

Total value of that equipment destroyed during the incident = $37,049

- 4 SCBA
- 16 bunker coats
- 16 bunker pants
- 13 pairs of boots
- 45 pairs of gloves
- 6 helmets
- 15 nomex hoods
- 3 hand lanterns
- 1 portable radio
- 1 monitor w/500 gpm nozzle
- 400 ft of 1 3/4 inch fire hose
- 200 ft of 3-inch fire hose
- 100 ft of 5-inch fire hose

APPENDIX D

Photographs

All photos were provided by Larry Wiles of the Nebraska City Fire Department.

#1 EnTire Recycling Facility November 2001, view of the south end of south building

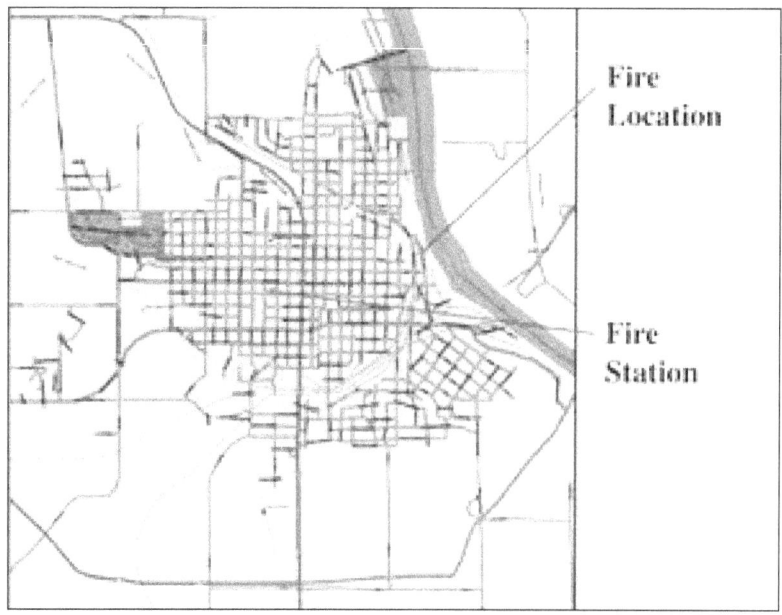

#2 EnTire location in relationship to fire station and Missouri River

Appendix D (continued)

Fire Ground Area

Power Plant:
  Command Post

Fire Area:
  North Building
  5 Silos
  South Building

Exposure Areas:
  Grain Elevator
  Barge Terminal
  Utilities Warehouse
  Elderly High Rise
  School

#3 Fire Ground Area and exposures

Appendix D (continued)

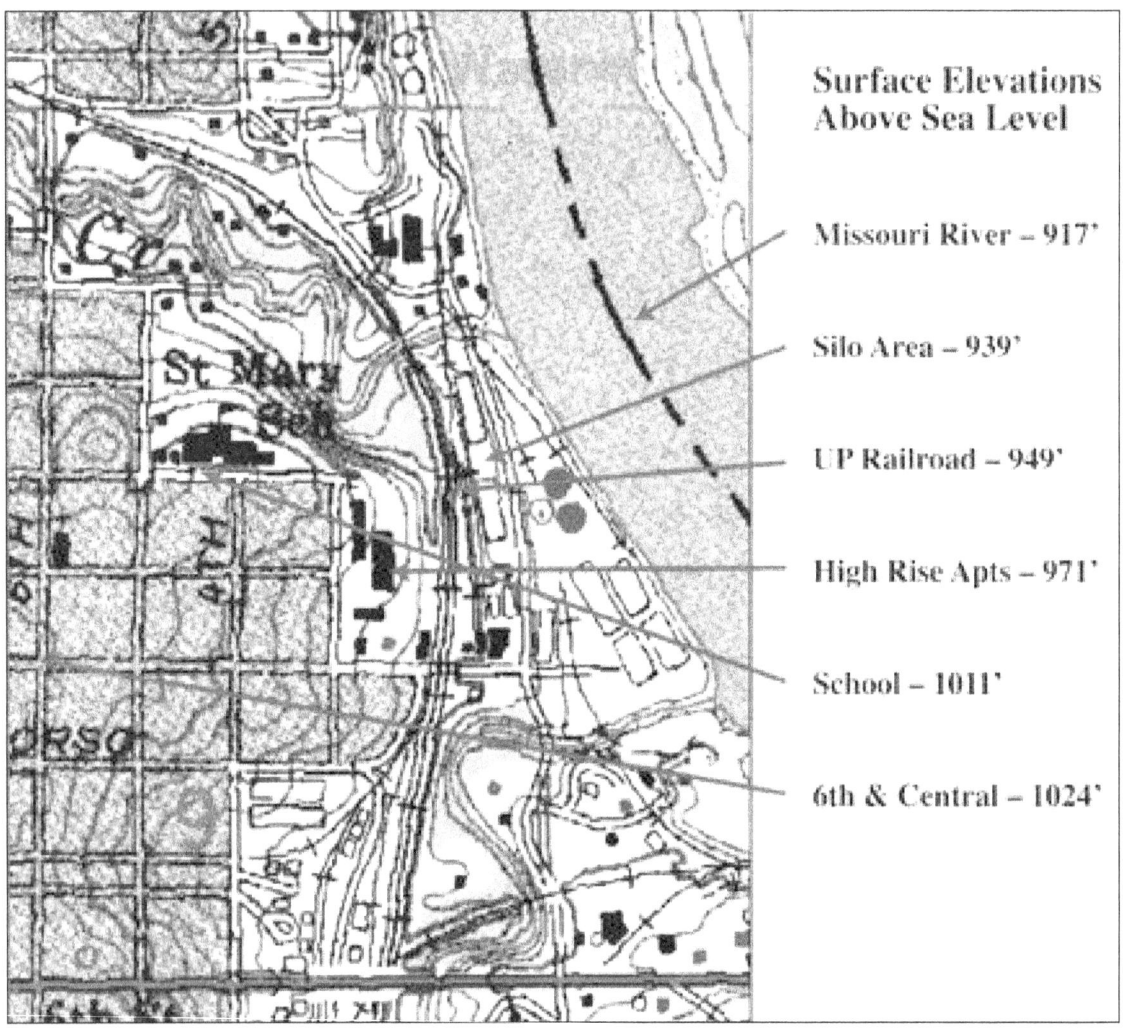

Surface Elevations Above Sea Level

Missouri River – 917'

Silo Area – 939'

UP Railroad – 949'

High Rise Apts – 971'

School – 1011'

6th & Central – 1024'

#4 Surface elevations above sea level

Appendix D (continued)

#5 Fire conditions 1.5 hrs after initial alarm, shows office, high tension wires, and SE & NE silos

#6 Two 8,000 gallon liquid nitrogen tanks, NE silo, smaller empty silo

Appendix D (continued)

#7 Fire conditions 1.5-2 hours after initial alarm, SE & NE silos

#8 Looking at west side of NW & SW silos and conveyor system
between north

Appendix D (continued)

#9 East side of SE silo 1.5-2 hours into incident

#10 Fire at top of elevator leg in center of four silos 3 hrs into incident

Appendix D (continued)

#11 Top of SE silo showing scaffolding and headhouse. Looking NW
just after sunrise

#12   11.5 hours into incident; headhouse has shifted to the south

Appendix D (continued)

#13 White ground fog is nitrogen escaping from
silo just minutes before explosion

#14 Photo taken just minutes before explosion

Appendix D (continued)

#15 Photo of chunks of crumb rubber thrown 300 to 400 ft by force of explosion

#16 Area of debris from explosion

Appendix D (continued)

#17 Chunks of crumb rubber clinging to side of railroad car

#18 Damage to engine

Appendix D (continued)

#19 Area where explosion occurred

#20 Damage to office

Appendix D (continued)

#21 Area of debris

#22 Fire erupts in north building on January 25 at 11:30 hours; view
looking NNW from south of the south building

Appendix D (continued)

#23 Looking NE from west side of silos

#24 North building fully charged with smoke

Appendix D (continued)

#25 North end of north building

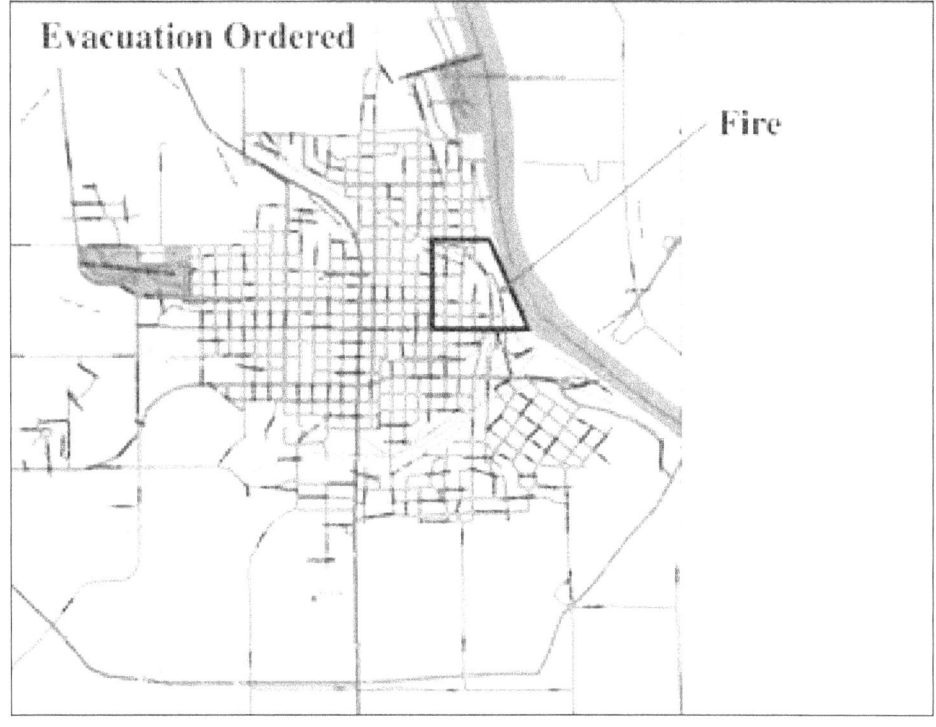

#26 Area evacuated due to fear of explosion; order lifted at 15:30

Appendix D (continued)

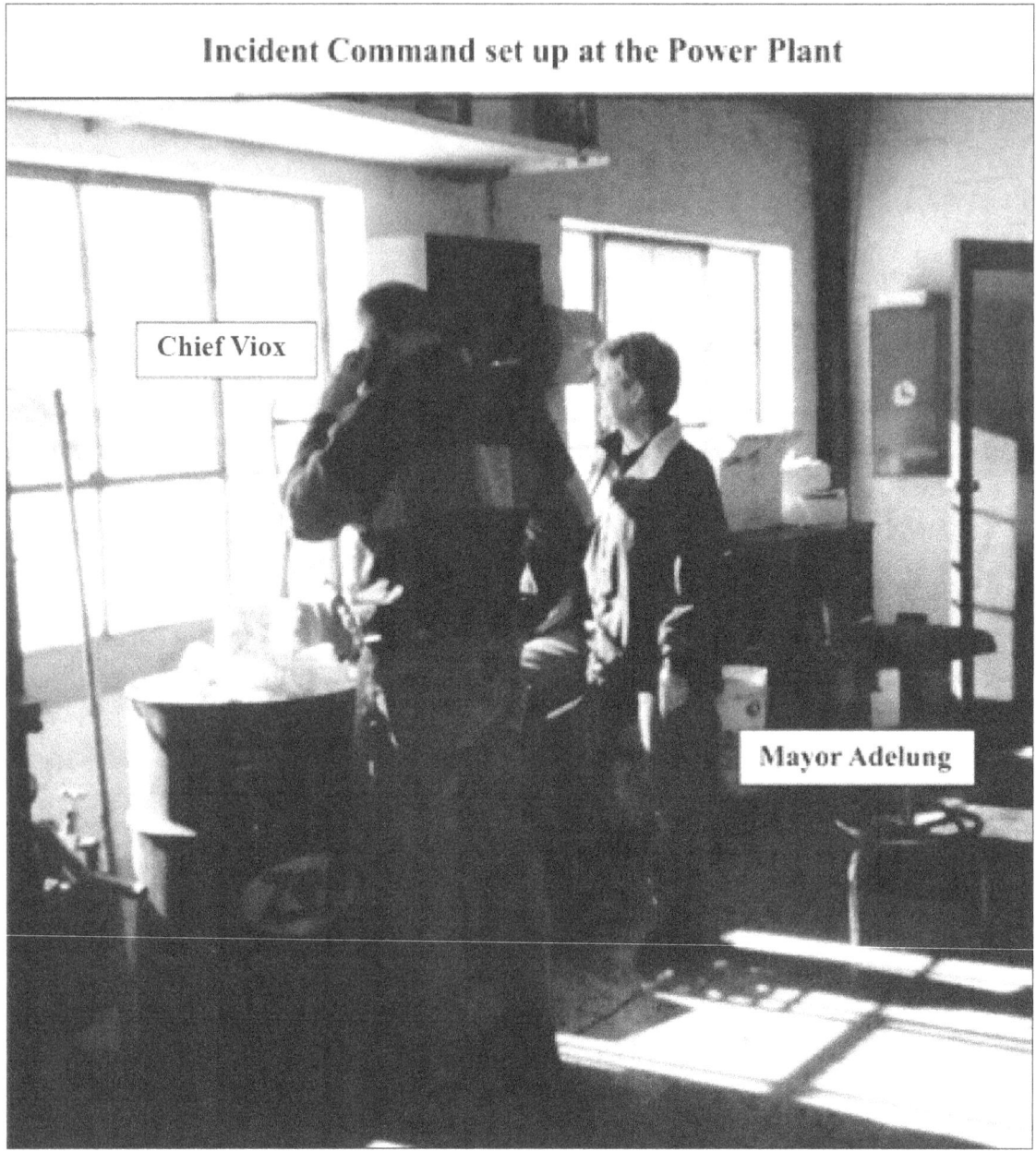

#27 Incident Command post in Municipal Power Plant

Appendix D (continued)

#28 Day 4, the 26th, dawns with little change

#29 Fire continues to smolder in two south silos

Appendix D (continued)

#30   2-1/2-inch foam nozzle from Williams Fire Control

#31 Williams Fire Control 2,500 gpm pump, one of two

Appendix D (continued)

#32 Pump wrapped with tarp and a construction heater to prevent freezing

#33 Headhouse begins to lean and silos shift on the 27th

Appendix D (continued)

#34 View from the south showing SW silo leaning into SE silo and scaffolding holding up headhouse

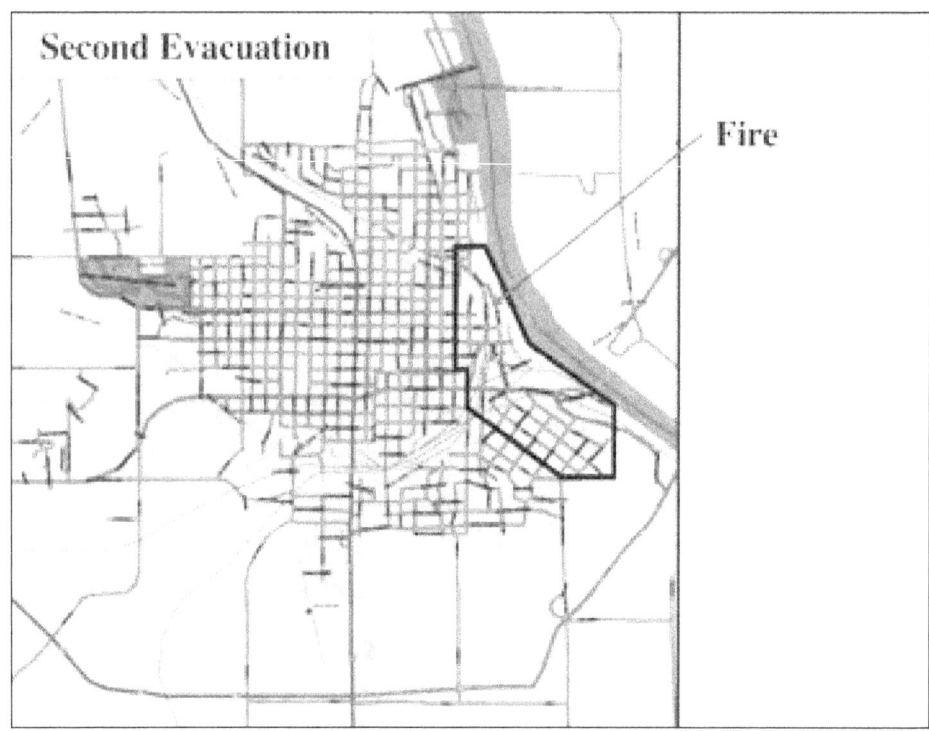

#35 Second evacuation area due to smoke

Appendix D (continued)

#36 On the 28th the headhouse falls resulting in SW and SE silo giving way and collapsing

#37 Fire intensifies after collapse; efforts concentrated on preventing fire from entering south building

Appendix D (continued)

#38 East side of south building removed and process begins to
remove material from south building in order to
spread it out and extinguish it

#39 Wreckers used to pull steel debris from fire

Appendix D (continued)

#40 Removing silo debris

#41 Spraying hose streams on smoldering rubber and debris to prevent spread as it was being removed

Appendix D (continued)

#42 Spraying hose streams on smoldering rubber and debris to
prevent spread as it was being removed

#43 Debris removal

Appendix D (continued)

#44 Reaching into NE silo to drag rubber out

#45 Pulling rubber out of NE silo

Appendix D (continued)

#46 Rubber is spread out to make sure fire is out prior to it being
transported to the landfill

#47 Opening up the last silo and final extinguishment

Appendix D (continued)

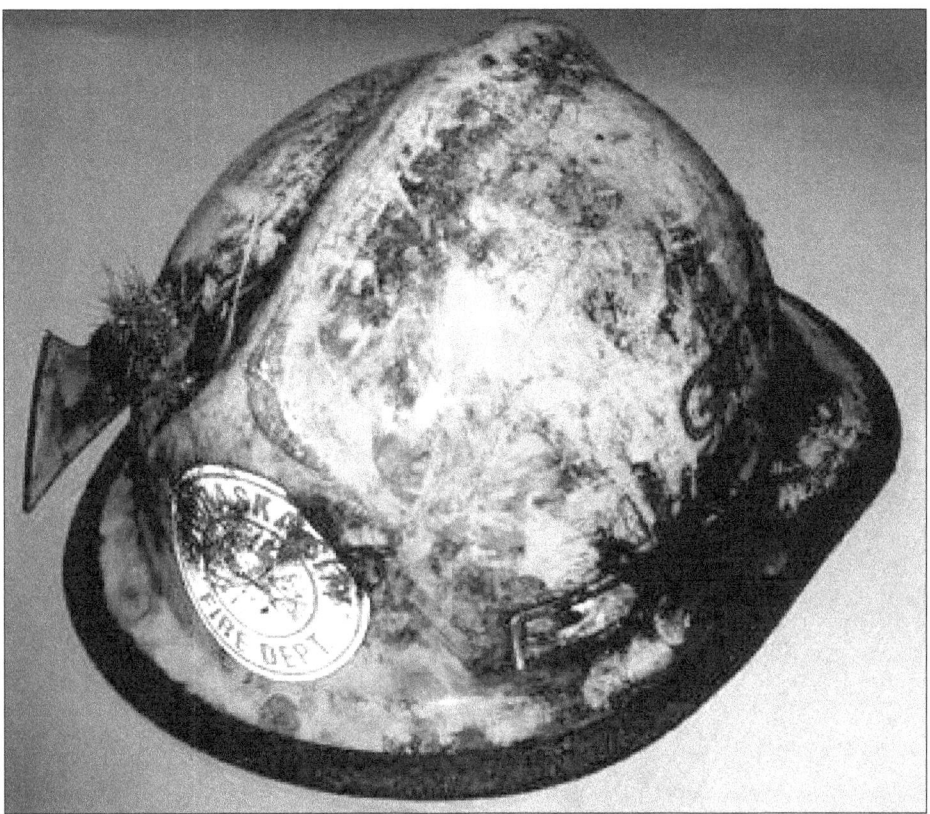

#48 Damage to fire helmet with chunk of rubber imbedded in shell

Appendix D (continued)

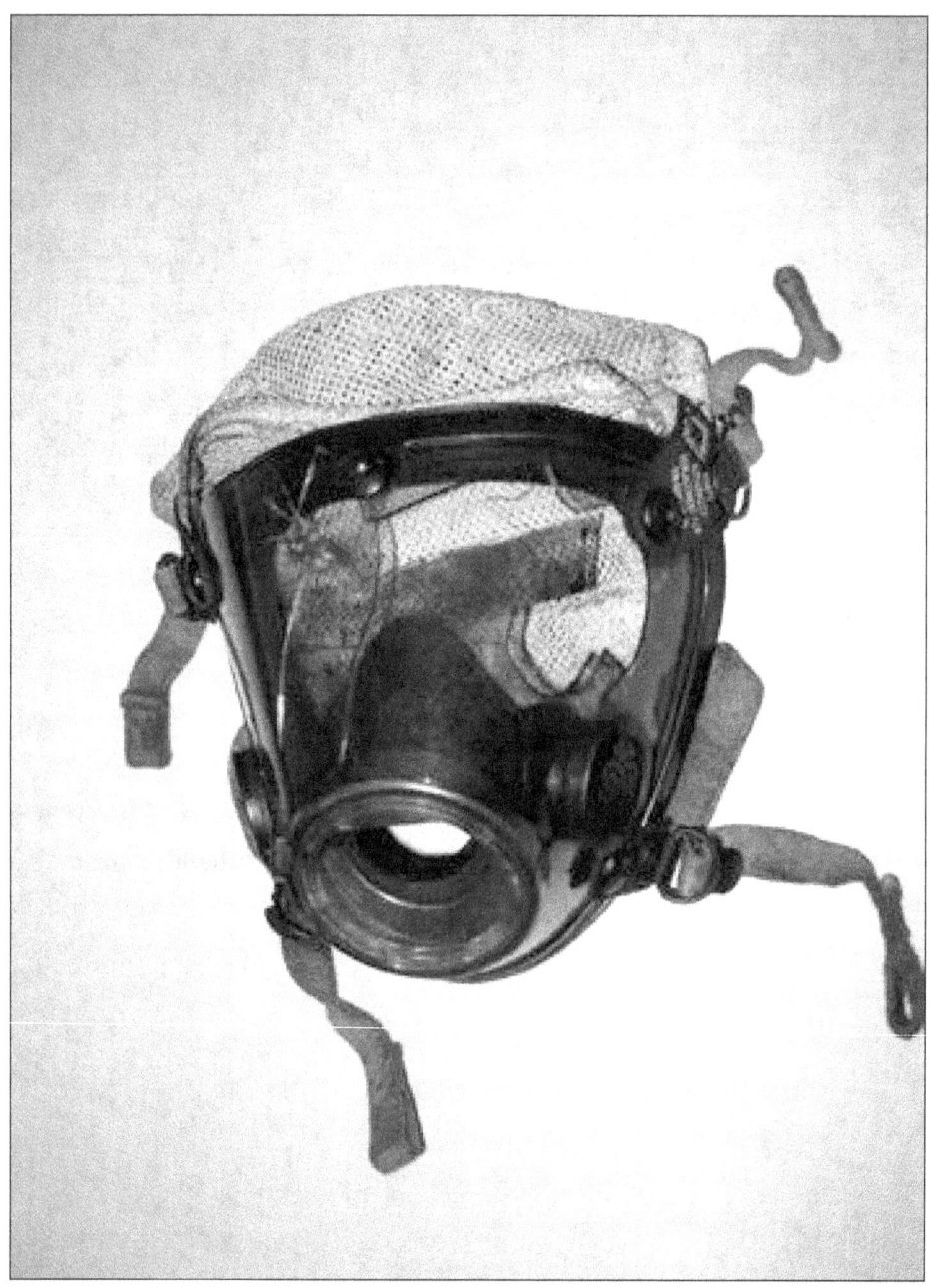

#49 Faceshield on SCBA cracked by flying debris from explosion

www.ingramcontent.com/pod-product-compliance
Lightning Source LLC
Chambersburg PA
CBHW081231170526
45165CB00009B/3039